THE CAVENDISH SQUARE ANIMAL LIBRARY

The World of BEARS

BY JUDY THORPE

Cavendish Square

Published in 2026 by Cavendish Square Publishing, LLC
2544 Clinton Street Buffalo, NY 14224

Website: cavendishsq.com

Disclaimer: Portions of this work were originally authored by Catherine Lukas and published as *Bears*.
All new material this edition authored by Judy Thorpe.

All websites were available and accurate when this book was sent to press.

Cataloging-in-Publication Data

Names: Thorpe, Judy.
Title: The world of bears / Judy Thorpe.
Description: Buffalo, NY : Cavendish Square Publishing, 2026. | Series: The Cavendish Square animal library | Includes glossary and index.
Identifiers: ISBN 9781502675095 (pbk.) | ISBN 9781502675101 (library bound) | ISBN 9781502675118 (ebook)
Subjects: LCSH: Bears–Juvenile literature.
Classification: LCC QL737.C27 T46 2026 | DDC 599.78–dc23

Editor: Caitlin McAneney
Designer: Deanna Lepovich

The photographs in this book are used by permission and through the courtesy of: Cover (image) Henk Bogaard/Shutterstock.com; cover and various pages (tree vector) cuong viet/Shutterstock.com; various pages (tape element) Tamara Luiza/Shutterstock.com; p. 4 Yatra4289/Shutterstock.com; p. 5 Maquiladora/Shutterstock.com; p. 6 Foreverhappy/Shutterstock.com; p. 8 (top) Sergii Chernov/Shutterstock.com; p. 8 (bottom) Menno Schaefer/Shutterstock.com; p. 10 (right) Malchev/Shutterstock.com; p. 10 (left) jo Crebbin/Shutterstock.com; p. 11 (top and bottom) elenabsl/Shutterstock.com; p. 12 ambquinn/Shutterstock.com; p. 13 KensCanning/Shutterstock.com; p. 15 Christopher MacDonald/Shutterstock.com; p. 16 Okyela/Shutterstock.com; p. 18 Robert Kniola/Shutterstock.com; pp. 20, 31 Bjorn H Stuedal/Shutterstock.com; p. 21 BlueRingMedia/Shutterstock.com; p. 22 Galyna Andrushko/Shutterstock.com; p. 23 Pavlo Baliukh/Shutterstock.com; p. 24 Andrii PIATNYCHKA/Shutterstock.com; pp. 26, 28 Sergey Uryadnikov/Shutterstock.com; p. 29 armminivet/Shutterstock.com; p. 30 udaix/Shutterstock.com; p. 32 GTW/Shutterstock.com; p. 34 AZ Septian/Shutterstock.com; p. 35 Sourabh Bharti/Shutterstock.com; p. 36 Lillian Tveit/Shutterstock.com; p. 37 Bigc Studio/Shutterstock.com; p. 38 Milton Rodriguez/Shutterstock.com; p. 39 WildMedia/Shutterstock.com; p. 40 MPH Photos/Shutterstock.com.

CPSIA compliance information: Batch #CSCSQ26: For further information contact Cavendish Square Publishing LLC at 1-877-980-4450.

Printed in the United States of America

CONTENTS

INTRODUCTION

Bears are incredible animals found throughout Asia, Europe, North America, and South America. Earth is home to eight different species of bears, all with their own amazing **adaptations**, behaviors, and special skills.

Bears are in the animal family *Ursidae*. They range from the smallest species, sun bears, to the largest species, polar bears. Between them in size, there are brown bears, black bears, sloth bears, spectacled bears,

Sun bears are the smallest kind of bear. They have superlong tongues to slurp insects!

This chart shows different bears and their relative sizes, from polar bears to the sun bear!

The giant panda is one of the rarest bear species on earth. There are fewer than 2,000 left in the wild.

Asiatic black bears (or moon bears), and giant pandas. While bear species such as polar bears and brown bears are sometimes shown as **aggressive** toward humans in movies and folktales, most bears just want to keep to themselves.

You can often identify bear species by their location and physical features. Polar bears live in the Arctic, in the most northern reaches of Canada, Alaska, Greenland, Norway, and Russia. Spectacled bears live in the jungles near the Andes Mountains in South America. It's not hard to tell these two bears apart. However, black bears and brown bears can be tricky to tell apart, because they share many of the same locations and fur colors. We have to look at other features to identify them.

Bears are able survive in many kinds of climates and conditions. Some bears shed their fur—called molting—once a year when their new fur grows in for the winter. Most bears that live in colder climates spend their winters hibernating, or in a deep sleep, which is an amazing adaptation that helps them survive harsh winters. Others have light, shaggy fur to live in warmer climates. Let's learn more about the life cycle, cool features, and habitats of bears from around the world.

Not all black bears are black, and not all brown bears are brown!

CHAPTER 1

BEAR BODIES AND FEATURES

The modern bears we know today first came from the Eurasian continent about 5 million years ago, though their **ancestors** had been **evolving** for millions of years before that. Over time, bears separated into different species based on their **environments**.

Bears have some great features that help them survive in these environments. Their size, fur type, and diets tell us a lot about how their bodies work for them, and how they play a **role** in their ecosystem.

BEAR BODIES

Bear bodies are built for survival. They come in different sizes. The sun bear, for example, is only four feet (1.2 m) long, weighing around 60 pounds (27 kg). It's great at climbing the many trees in its environment. However, a brown bear can grow to almost 10 feet (3 m) long and weigh up to 1,500 pounds (680 kg).

All bears have **nonretractable** claws and short tails. Their legs are relatively short compared to the rest of their body. Because of their size, larger bears rarely have to worry about predators.

BIGFOOT OR BIG BEAR?

Bears are often some of the largest animals in their environment. They can stand pretty tall—especially when they are upright on two legs. Some scientists think that they may even be the answer to Bigfoot sightings! Bigfoot is a mythical creature that looks huge, hairy, and human-like. It's never been caught or recorded in any scientific way.

Some scientists looked into reports of Bigfoot sightings in the Pacific Northwest, and mapped them out alongside bear populations. They found areas with higher bear populations have more Bigfoot sightings. Therefore, they believe that most Bigfoot sightings are black bears.

Do you think you could tell the difference between a standing black bear and Bigfoot, especially in the dark, dense forest?

Check out the differences between a black bear and a grizzly bear.

Bears come in many different colors, and those colors can sometimes help you identify what species it is. The most common types of bears are black bears and brown bears. It isn't always easy to tell them apart. Black bears can be black, brown, reddish, tan, or even white. Some have a patch of white fur on their chests. Brown bears, including grizzly bears, are usually brown, but the color of their fur ranges from tan to black.

So how do you tell the difference between black and brown bears? One way is by their size. Brown bears are bigger and have a hump on their shoulders. Their ears are short, round, and smaller than the ears of

THAT'S WILD!

The Kermode bear lives off the coast of British Columbia in Canada.

Grizzly bear fur is light at the end of each hair, which gives these bears a "grizzled" appearance.

black bears. A brown bear's claws are longer and a lighter color than those of black bears.

One rare kind of black bear has white or cream-colored fur. It is not a polar bear. Native Americans named this bear "spirit bear." It is also known as a Kermode bear. There are between 400 and 1,000 Kermode bears in the wild today.

A bear's fur color can also tell you about its environment. Polar bears appear to have white fur, which helps them blend into their icy, snowy environment. They have a double-layered coat to keep them warm in the Arctic. Panda bears also have

BEARS IN REAL LIFE

Many people who see bears only in movies and TV shows think that they are often aggressive. It's true that mother bears may protect their cubs by becoming aggressive. However, most bears just want to be left alone. The black bear is the least aggressive bear species.

Most movies also show bears growling as a threat. However, bears don't really make growling sounds! Most bears are actually pretty quiet. They do make noises to communicate, using their "voices" to show their emotions. Cubs whine or cry when they are upset or "chuckle" when they are happy. Adults may grunt when friendly or make a blowing noise when afraid.

This bear has white fur, but it's not a polar bear! It's the Kermode bear. You can look at its size and habitat to help you identify it.

THAT'S WILD!

Polar bear skin is actually black, and their fur is not really white. The thick fur is translucent, or clear, and the hairs are hollow, so it looks white.

white fur to **camouflage** in snowy surroundings, but also black markings to help it hide in shady spots.

BEAR ABILITIES

Bears use their unique physical features to carry out the activities that help them stay alive. Bears are most active in the early morning and evening. During summer and early fall, bears may be active all day. They spend a lot of time looking for food, a mate, or a place to hibernate.

Bears usually walk on all fours, but they can stand on just their back feet too. Bears stand up to reach food, see into the distance, fight off an attacker, or sniff the air. Bears may look slow because of their large size, but they can run as fast as 30 miles (48 km) per hour for short distances. That's faster than the fastest human **sprinter** and is the maximum speed cars are allowed to go in many neighborhoods across the United States!

THAT'S WILD!
Black bears can swim 1 mile (1.6 km) at a time without stopping.

Bears are good swimmers. They go into the water to catch fish, cool off, escape pesky insects or a predator, and sometimes just for fun. Polar bears are the best bear swimmers. They can swim for hours in the icy waters of the Arctic. The powerful bodies of bears help them swim for long periods of time.

Having strong front and back legs also helps bears climb. Black bears learn to climb trees when they are young. Trees offer food (fruit, sweet sap, and insects), safety, and branches for resting. Some bears hibernate in the crook of a branch for the whole winter! A black bear climbs by digging into a tree trunk with its front paws and pushing up with its hind legs. It climbs down backward, hind legs first, the way humans do. Bears also slide down tree trunks or jump from low branches.

This black bear cub climbs a tree.

Bears have curved claws that allow them to turn over rocks and logs, pick berries from bushes, catch fish, and dig up roots.

Most bears are curious and intelligent creatures. Bears investigate new objects, noises, and smells. They have even been known to outsmart humans who try to keep them away from food in garbage cans and at campsites! They also have good memories. They can find their way back to places where they have found food—even places they have visited only once before. They remember paths and routes they've walked in the past.

BEAR SENSES

A bear's senses also help them survive in the wild. Most bears have good eyesight during the day. However, their night vision is even better, aided by a reflective lining in the back of their eyeball. They can detect movement in their surroundings well, which helps keep them safe and find food. Bears have great hearing as well, and they can hear in all directions. Their hearing is twice as sensitive as that of a human.

> **THAT'S WILD!**
> A bear's sense of smell is 2,100 times stronger than a human's!

Of all the senses, a bear's sense of smell is the strongest. Bears can sniff out tiny insects inside logs or a dead animal more than a mile away! A bear's sense of smell is seven times better than that of a bloodhound, which is often considered the kind of dog with the best sense of smell.

Bears also use their sense of smell to communicate with each other. They leave their scent on trees and bushes to "tell" other bears they are in the area. Bear cubs know the scent of their mother, which helps keep them safe. They can find their mother by scent when they are lost, in danger, or hungry and need milk.

Giant pandas have a meat eater's digestive system, but eat mostly bamboo. They are one of nature's most interesting vegetarians!

CHAPTER 2

BEAR BEHAVIOR

What do bears do all day? And what do they get up to at night? Bears can be active at any time of day or night, depending on the time of year and their habitat. Those that live in temperate or colder climates spend a lot of time looking for food during the warmer months. Then, they get a lot of rest during colder months. Let's explore more about a bear's life.

FINDING FOOD

During warmer months, bears spend most of the day eating. Grizzly bears can eat up to 30 pounds (14 kg) of food a day! This will last them through the winter. What are their favorite foods?

The type of food that bears eat varies, depending on where they live and the time of year. Most bears are omnivores, which means they eat plants and animals. Brown bears eat practically anything, but their favorite foods are fruit, nuts, acorns, and insects. They also eat fish, small mammals, and honey. In some areas, brown bears hunt larger animals such as moose, elk, and mountain goats.

THAT'S WILD!

Giant pandas eat mostly bamboo. An adult can eat up to 84 pounds (38 kg) of bamboo per day!

Bears use their teeth and claws to get their food. A bear's teeth are shaped for eating both plants and animals. The pointed front teeth hold food, such as a wriggly fish, while the bear's flat molars work like a human's for crushing and grinding. With their big paws, bears dig into logs and overturn rocks to grab bugs and other critters to eat.

BEARS ON THE HUNT

While many bears find and eat plants as a big part of their diet, polar bears are carnivores, which means they eat only meat. Living in the Arctic, there's not a lot of plant life to eat! Polar bears spend a lot of their time hunting. They like to eat ringed seals, which are ocean mammals with enough fat to give polar bears the energy they need. Polar bears will also hunt walruses and eat the bodies of whales that are already dead.

To catch a seal, polar bears wait on ice for the seal to come to the surface of the water. Then the polar bear will grab the seal and bring it to the ice to eat it. Nothing eats a polar bear, so they are apex predators—the top of the food chain!

This polar bear teaches her cub how to hunt for food.

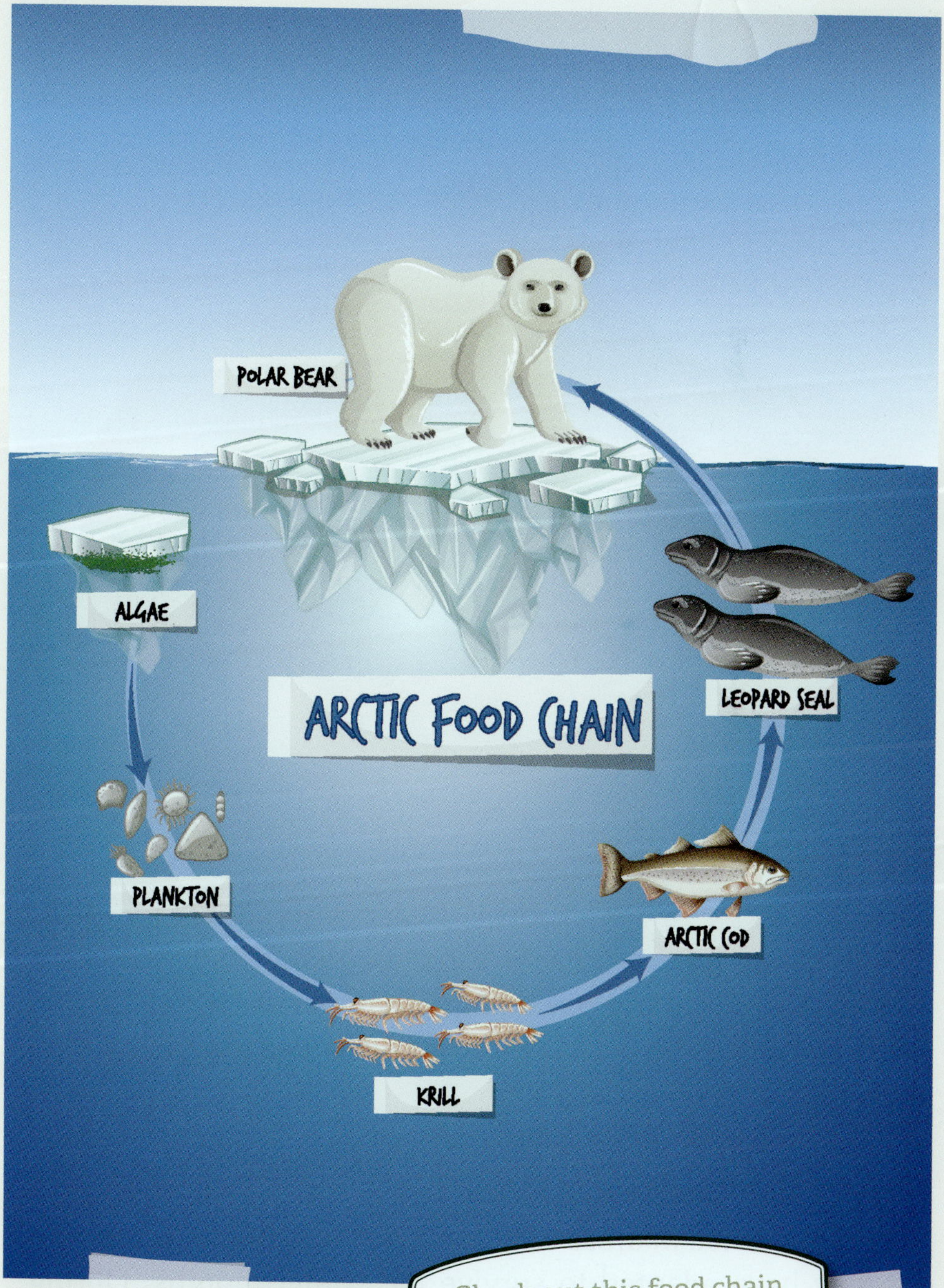

Check out this food chain showing polar bears at the very top!

Large adult males get their first pick of the salmon. Then, mothers and cubs can fish too.

Brown bears also hunt for their favorite food—salmon. They use their big paws to grab salmon as they swim and jump up waterfalls. A male brown bear can eat more than 30 salmon a day!

WINTER ADAPTATIONS

Not all bears live in areas with cold weather. But for those that do, they have adapted to survive the winter months. During late summer and early fall, bears gain weight to get ready for winter. They eat as much as they can, day and night. Depending on the amount of food available, black bears at this time of year may gain six pounds (2.7 kg) a day. The extra fat keeps the bears alive during winter and helps keep them warm. As cold weather approaches, bears stop eating and get sleepy. They are preparing to enter a state commonly known as hibernation, though some scientists refer to what bears experience as torpor.

In cold climates, bears sleep in their dens for as long as seven months—usually starting in mid-September. The colder the winter, the longer bears sleep. They do not eat, drink, or

THAT'S WILD!

Some scientists believe that bears are not true hibernators. Their body temperatures and heart rates do not drop greatly like those of other animals, such as ground squirrels and chipmunks.

If you ever see a big hole in the forest floor—it might be a bear den!

WHERE DO BEARS HIBERNATE?

Bears hibernate in dens, which they begin to get ready in September or October. Some bears use the same den every winter. Others choose a new one every year.

Bears choose many kinds of places to build their dens, including rock caves. Rock caves provide protection from harsh weather conditions. Bears may also use their strong paws to dig holes in the ground. Sometimes they make their dens high up in a tree. Bears that live around people may even make their den underneath porches. Bears may need to leave their dens in winter if the den floods or if something happens to make them feel unsafe.

Bears choose safe places to sleep, such as caves like the one shown here.

THAT'S WILD!

Bears in zoos may not hibernate at all if they're being fed regularly. However, some zoos allow their bears to go through natural hibernation.

get rid of waste during this time. A bear that sleeps through winter needs only half the amount of oxygen it needs when it's awake and active. Sometimes bears wake up and come out of their dens, but they come back and go to sleep again.

During hibernation, bears live off the body fat they stored before winter. When the weather warms up, usually in April or May, the bears come out of their dens. At first, they are weak. They have lost as much as half their body weight. But as plants grow, the bears begin to eat and gain weight again. Then their energy returns. It's time to eat as much as possible once again to get ready for the next winter!

This baby brown bear depends on its mother for food and protection.

CHAPTER 3

A BEAR'S LIFE

A bear's day-to-day life depends on where it lives, what food is available, and what challenges it may come across. But all bears have something in common. They begin as a tiny, helpless cub, and over time, they grow to be many times that size. Explore the life cycle of a bear to discover the journey each one takes as they grow.

A BABY'S BEGINNINGS

When a pregnant bear starts to hibernate, the baby bear or bears inside her grow quickly. The bear cubs are usually born in January or February, while the mother is hibernating. A mother bear usually gives birth to one to three cubs at a time, but scientists have recorded as many as six born to one mother.

At birth, baby bears are tiny, toothless, and covered with wispy fur. They cannot see, hear, or smell until they are older. Baby bears are able to walk when they are five or six weeks old.

THAT'S WILD!

Bears are usually solitary, or live alone, except for mothers and their cubs.

When a mother bear gives birth, she licks her cubs and keeps them close to her. Once they begin to nurse, or drink milk from her body, she falls back into her deep sleep. The cubs eat and sleep for two to three months. The mother bear's milk is high in fat, and that helps the babies grow quickly. By spring, they're ready to greet the world outside their den.

GROWING UP

Even as baby bears enter the world, they still need their mothers for protection, and then, for education. Their mother teaches them to find food. She shows them how to climb trees and rescues them from any danger by carrying them in her mouth. She also shows them how to escape from predators. Mother bears are very protective of their little ones and may attack if their cubs are in danger.

Bear cubs are playful. They climb trees, wrestle, play fight, and chase each other. Playing helps make them strong. Playing also gives cubs the social skills they need. By learning to understand

THAT'S WILD!

A mother bear may recognize her offspring for years and allow them to search for food in her territory.

Baby bears are great at climbing trees!

HELPLESS BABY PANDAS

Baby pandas are totally helpless. They are some of the smallest newborns, in comparison to their mother's size, of any mammal on Earth. Their mother weighs 900 times more than them when they are born. The reason baby pandas are so underdeveloped may be because the bamboo that mother pandas eat doesn't have enough nutrients to keep their pregnancy going. Baby pandas are unable to crawl, see, hear, or even poop on their own! Their mothers keep them close, feeding them fat-rich milk. They are born naked, but soon baby pandas grow white fur as well as their black markings.

THAT'S WILD!

When a mother panda has two babies at once, she usually leaves one to die because it's too much work to feed more than one baby around the clock.

This tiny baby panda looks nothing like its giant mother!

other bears' behavior, cubs will be ready to know if a strange bear is friendly or not.

In fall of the cubs' first year, mother and cubs prepare a den. They sleep together until spring. Not long after the cubs leave the den for the second time, when they are about 17 months old, they are ready to leave their mother. They will now be solitary animals, until they make their own families.

THE CYCLE BEGINS AGAIN

Once a female bear is about five years old, she is ready to mate. Then, the life cycle begins again. Bears often mate in

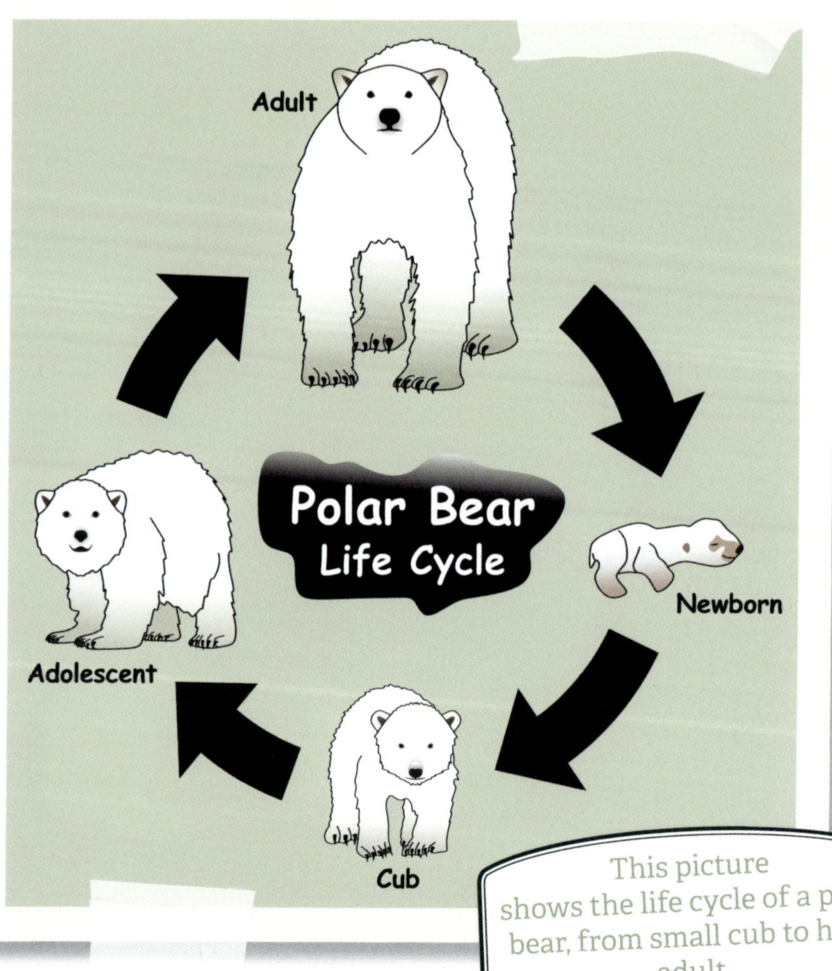

This picture shows the life cycle of a polar bear, from small cub to huge adult.

the summertime, but babies don't really grow much inside the mother until she hibernates. Mother black bears give birth to cubs every two or three years. Male bears may mate with many females in a season, and they don't have anything to do with caring for the babies.

It takes a lot of energy for a polar bear to get enough food in the Arctic. Because of that, their life span isn't as long as other bear species.

Other than humans, bears have few threats in the wild. Mountain lions, crocodiles, and snow leopards may attack an adult bear, but more often, they go after the cubs. Because of this, bears have a relatively long life cycle in the wild—up to 25 years. In **captivity**, they can live to be 50 years old!

Polar bears have a thick layer of **blubber** under two coats of fur. That helps **insulate** their bodies from the **extreme** cold.

CHAPTER 4

BEARS AT HOME

Why are polar bears so different from sun bears? Why are grizzly bears so different from giant pandas? The differences between bear species often have to do with their environment. They have evolved over time to be masters of their own ecosystems, living off the food that is available and gaining the skills that help them survive. Let's take a look at bears in their home habitats!

AT HOME IN THE ARCTIC

The Arctic is an icy, cold region in the northernmost part of Earth. Polar bears are some of the few large land animals that can survive in the coldest parts of the Arctic. Temperatures can get as low as -70 degrees Fahrenheit (-57 degrees Celsius)!

While polar bears don't have to worry about predators, they do have to worry about finding enough food. Sea ice is melting due to climate change, so polar bears have more trouble hunting seals from the sea ice.

THAT'S WILD!

Sometimes, grizzlies and polar bears mate, which causes a bear **hybrid** called a "pizzly."

Sometimes, brown bears and polar bears live in the same

places, such as parts of northern Canada and Alaska. This is happening more and more as climate change happens. Grizzlies move north and polar bears move south in search of food.

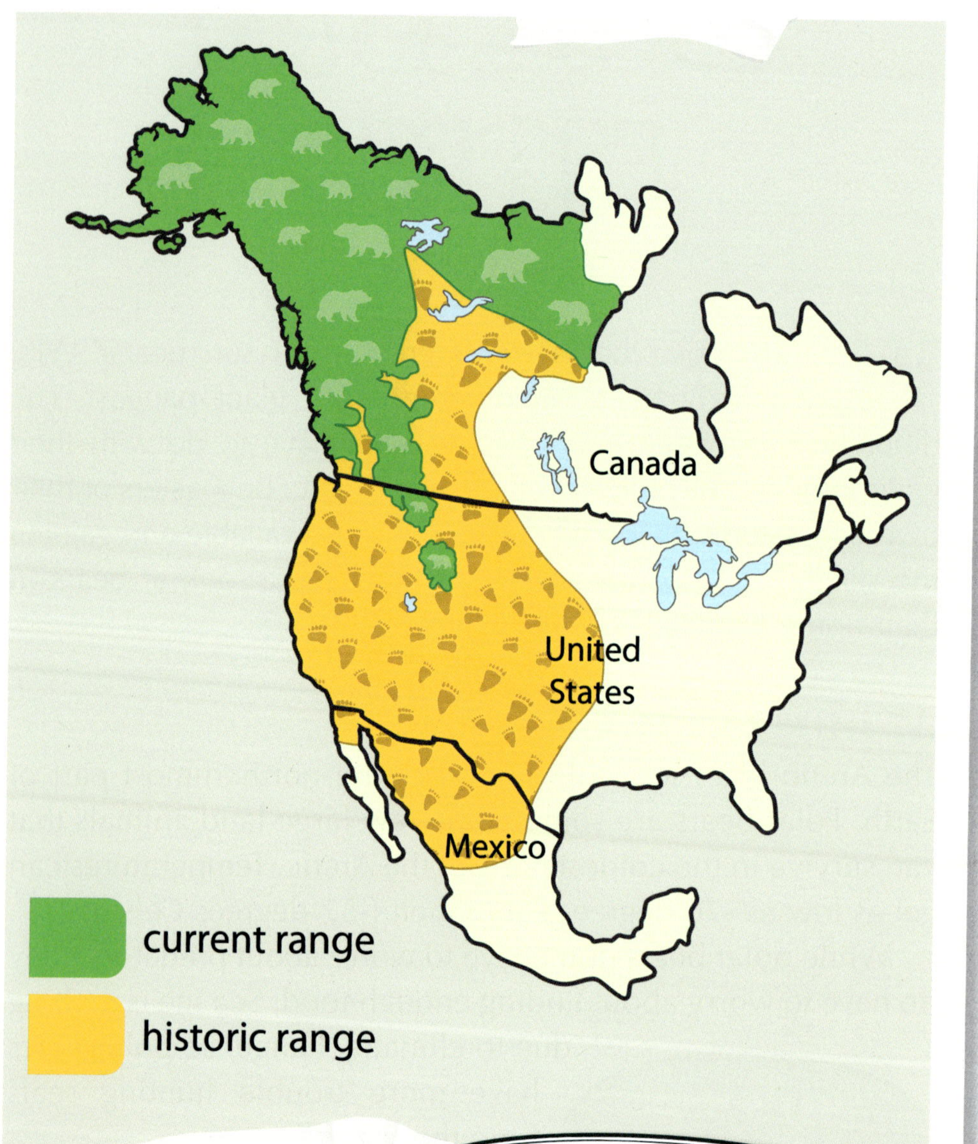

This map shows how the grizzly bear range has changed over time.

AT HOME IN THE TROPICS

Not all bears live in cold places with long winters. Sloth bears, sun bears, and spectacled bears are at home in tropical and subtropical parts of Earth. The forests here have a warm, wet climate and lots of trees. There are plenty of bugs, fruits, and leaves to eat. Because of that, these species are omnivores that are great at climbing trees.

Sloth bears live in tropical and subtropical forests in Sri Lanka, Nepal, and India. They are slow-moving, noisy bears that live on insects and fruit. They use their large claws to dig into

THAT'S WILD!

Sloth bears have a lot of forest predators to worry about! Tigers, wild dogs, and leopards may attack them.

Sloth bears do not have an undercoat so they can stay cool in tropical weather.

rock-hard termite mounds and gobble up the insects inside. They also climb trees, knocking down beehives to get to the honey inside.

Sun bears live in the forests of Southeast Asia. They spend a lot of their time in the trees, even sleeping in the branches. Like sloth bears, they mostly eat fruits and insects, and their favorite food is honey. Since they often get into beehives, they have thick skin and fur to protect them from stings. Sun bears can

This sun bear is relaxing high up in a tree.

ASIATIC BLACK BEARS

Asiatic black bears live in the thick forests and mountain areas of Asian countries such as China and Thailand. Some of them live in tropical rainforest environments. Others live in more temperate forests, where it gets colder in the winter months. In these places, Asiatic black bears may hibernate, just like their black bear and brown bear cousins.

Asiatic black bears have longer, thicker hair around their necks, which makes them look like they have a mane, somewhat like a lion. Their fur is mostly black or brown, but they have a lighter moon-shaped mark on their chest. That's how they got the nickname "moon bear."

Asiatic black bears are threatened by the loss of trees in their habitats.

THAT'S WILD!

In some Asian countries, Asiatic black bears are poached, or hunted, for their body parts, which are used in some traditional medicines.

rip open beehives with their long claws and then get the honey out with their long tongues!

The only bear species in South America is the spectacled bear. It lives in the cloud forests of the Andes Mountains. Cloud forests are thick tropical rainforests along the steep slopes of mountains. Spectacled bears

Spectacled bears get their name from the markings around their eyes that look like spectacles, or glasses.

HABITATS AT RISK

Bears are wild animals that depend on their habitats for survival. Many bear species are at risk of dying out due to the loss of their habitats, especially the giant panda. Polar bears are at risk of losing the sea ice that they depend on for hunting. This is a big problem because of climate change. The Arctic is heating up in ways that polar bears aren't suited for.

Bears in forested areas are suffering from deforestation, or the cutting down of too many trees for building or lumber. When trees are cut down, it makes it hard for bears to find enough food or a place to make their dens. They might enter human communities and be harmed or killed out of fear. It's important that bears around the world have the habitats they need to survive.

Cutting down trees in a bear's habitat may push them into new territories, which can harm ecosystems and scare communities.

can live as high up as 14,000 feet (4,267 m) in elevation. Like their other tropical bear cousins, they spend a lot of time in trees, finding fruits to eat.

AT HOME IN NORTHERN FORESTS

Black bears and brown bears are both found in the forests of the Northern Hemisphere, including in North America. These bear species are the ones that most often hibernate to survive the coldest months of the year. Their homes have seasons with very different weather: warm summers and cold winters.

Brown bears and black bears are both known to steal food from human campsites—so make sure to keep your food locked up while camping!

Black bears have a wide range—throughout North America and even into northern Mexico. They live in many different habitats, such as mountains and swamps, but are most at home in forests. They eat whatever they can find, including salmon, roots, bugs, and berries. They'll eat almost anything so they can bulk up before winter.

Brown bears have the largest range of any bear species, found in the temperate northern forests of North America, Asia, and Europe. Kodiak bears and grizzly bears are both kinds of brown bears. Like black bears, brown bears will eat almost anything to prepare for winter. Some are lucky enough to have access to huge amounts of migrating salmon. Others eat small animals, bugs, fruits, and **carrion**.

As you have learned, the world of bears is extremely diverse. From the polar bears of the Arctic and the brown bears of North America to the sun bears of Southeast Asia, bears make their homes in a variety of biomes and have adapted to life in each of these unique, or special, environments. Learning about these powerful creatures helps us understand why they are so important in their ecosystems and why they must be protected.

THINK IT THROUGH

1. Why do you think some species of bears have two coats of fur, while others have one?
2. What could be a problem with polar bears and grizzlies sharing the same range?
3. Why do you think bears have often played the role of villain in folktales?
4. Why is it especially important to keep your distance from a mother bear and her cubs?
5. Why might it be important for bears in captivity to be able to hibernate as they would in nature?

GLOSSARY

adaptation: A change in a body that makes it better suited to certain conditions.

aggressive: Showing a readiness to attack.

ancestor: One of the living things from whom another is descended.

blubber: A thick layer of fat on marine mammals that is meant to keep these animals warm.

camouflage: Colors or shapes on animals that allow them to blend in with their surroundings.

captivity: The condition of not being free.

carrion: The flesh of a dead animal.

environment: The conditions that surround a living thing and affect the way it lives.

evolve: To grow and change over time.

extreme: More than what is expected.

hybrid: The offspring of two animals or plants of different kinds.

insulate: To keep heat from escaping a body or object.

nonretractable: Unable to be drawn back in.

sprinter: One who runs at high speeds for short periods of time in races.

role: A part something plays in a larger system.

FIND OUT MORE

Books

Backhouse, Frances. *Grizzly Bears: Guardians of the Wilderness*. Victoria, BC: Orca Book Publishers, 2023.

Brydon, Alli. *A Bear's Tale*. New York, NY: DK Publishing, 2024.

Gish, Melissa. *Black Bears*. Mankato, MN: Creative Education and Creative Paperbacks, 2024.

Roggio, Sarah. *Polar Bears vs. Seals: Food Chain Fights*. Minneapolis, MN: Lerner Publications, 2025.

Rose, Rachel. *Sun Bear*. Minneapolis, MN: Bearport Publishing Company, 2024.

Websites

Bears

www.nationalgeographic.com/animals/mammals/facts/bears-grizzly-polar-panda

Learn more about the different bear species with *National Geographic*.

Brown Bear

animals.sandiegozoo.org/animals/brown-bear

Explore information about brown bears, from their adaptations to how they hibernate.

Giant Panda

kids.nationalgeographic.com/animals/mammals/facts/giant-panda

Enjoy amazing facts about one of the rarest bears on Earth—the giant panda.

Sloth Bear

nationalzoo.si.edu/animals/sloth-bear

Learn more about the sloth bear with the Smithsonian National Zoo.

Sun Bear

www.britannica.com/animal/sun-bear

Discover fascinating facts about the smallest bear species—the sun bear.

Publisher's note to educators and parents: Our editors have carefully reviewed these websites to ensure that they are suitable for students. Many websites change frequently, however, and we cannot guarantee that a site's future contents will continue to meet our high standards of quality and educational value. Be advised that students should be closely supervised whenever they access the Internet.

INDEX

ABOUT THE AUTHOR

Judy Thorpe studied anthropology at Imperial College London. Through her courses, she developed a keen interest in exploring the ways various cultures' myths and politics impact each other. She is also an animal lover and enjoys learning as much about various animals as possible. When she is not writing, she enjoys rock climbing or curling up with a book and her cat, Punch.